Zeta and Eta Functions
A new hypothesis

Ashwani Kumar Thukral

DEDICATION

To the memory of my elder brother
JITENDRA PAL THUKRAL

.

CONTENTS

Preface

Zeta and eta functions have intrigued me for several years as to why these can not be defined for negative numbers without analytic continuation. I am presenting in this manuscript my own perception of these functions. Though I am a biologist by education and profession, with no formal education in mathematics, it is my endeavour to understand the basics of the subject to which I am least exposed. I am very clear in my mind that creation of knowledge is not out of bounds for any one. Sometimes small ideas by amateurs may also be sustainable for further development of knowledge. This is with this intention that I have presented my hypothesis in this book. I am not a number theorist but just trying to understand the numbers which fascinate all. In this hypothesis I have tried to define the zeta and eta functions of negative numbers, without analytic continuation.

Ashwani K. Thukral

Professor,
Department of Botanical and Environmental Science
Guru Nanak Dev University
Amritsar – 143005 (Punjab) India.
email: akthukral.gndu@gmail.com

i

Acknowledgements

This is to acknowledge the academic interaction with the teachers and students of Guru Nanak Dev University, Amritsar – 143005 (Punjab) India. My wife and son have always been a source of inspiration to me for my professional growth.

1

INTRODUCTION

1.1 Euler-Riemann's zeta function

Zeta function, $\varsigma(x)$, is one of the most important functions in mathematics having applications in various fields [2-7,10,12,14,18]. Zeta function of a real number $x > 1$, is defined as sum of infinite harmonic series of natural numbers:

$$\varsigma(x) = \frac{1}{1^x} + \frac{1}{2^x} + \frac{1}{3^x} + ..., \ x > 1,$$

$$= \sum_{n=1}^{\infty} \frac{1}{n^x}, \ x > 1. \tag{1}$$

In 1731, Euler gave a method [1,5,13] for the computation of the value of $\varsigma(2)$ .In 1749, Euler gave functional equation for analytic continuation of the zeta function for $x > 0$ [21],

$$\varsigma(x) = \frac{1}{\left(1 - 2^{1-x}\right)} \sum_{n=1}^{\infty} \frac{(-1)^{n-1}}{n^x}, x > 0, x \neq 1,$$

$$= \frac{1}{\left(1 - 2^{1-x}\right)} \eta(x), x > 0, x \neq 1. \tag{2}$$

where $\eta(x)$ is the Dirichlet eta function. Euler proved identity relating the zeta function to the product of all primes (p) [8,9],

1

$$\varsigma(x) = \prod_p \left(1 - \frac{1}{p^x}\right)^{-1}, x > 1,$$

and found the values of $\varsigma(2), \varsigma(4), \varsigma(6)$ etc. Bernhard Riemann in the year 1859 extended the Euler's zeta function for real numbers (Eqn. 1) to complex numbers (s),

$$\varsigma(s) = \sum_{n=1}^{\infty} \frac{1}{n^s}, \operatorname{Re} s > 1,$$

where $s = x + iy$, x and y are real numbers. The integral for Riemann's zeta function is

$$\varsigma(s) = \frac{1}{\Gamma(s)} \int_0^{\infty} \frac{x^{s-1}}{e^t - 1} dt, \operatorname{Re} s > 1. \tag{3}$$

Series representation of Reimann's zeta function is [19,24],

$$\varsigma(s) = \frac{1}{1 - 2^{1-s}} \sum_{n=1}^{\infty} \frac{(-1)^{n-1}}{n^s}, \operatorname{Re} s > 0, s \neq 1.$$

The integral form is

$$\varsigma(s) = \frac{1}{\left(1 - 2^{1-s}\right)\Gamma(s)} \int_0^{\infty} \frac{t^{s-1}}{e^t + 1} dt, \operatorname{Re} s > 0, s \neq 1. \tag{4}$$

Functional equation of zeta function of real negative numbers is:

$$\varsigma(1-s) = 2(2\pi)^{-s} \cos(\pi s / 2) \Gamma(s) \varsigma(s), \operatorname{Re} s > 0.$$

For negative integers, the zeta function is related to Bernoulli numbers (B_n) by the following equation [21],

$$\varsigma(-n) = -\frac{B_{n+1}}{n+1}, n \geq 1.$$

Figures (1,2) represent graphs for Riemann's zeta function.

Table 1 gives the values of Riemann's zeta function for some negative real numbers.

1.2 Dirichlet's eta function

Dirichlet eta function is an alternating zeta function [20]. Dirichlet series for real positive numbers is give n as

$$\eta(x) = \frac{1}{1^x} - \frac{1}{2^x} + \frac{1}{3^x} - ..., x > 0,$$

$$= \sum_{n=1}^{\infty} \frac{(-1)^{n-1}}{n^x}, \quad x > 0. \tag{5}$$

The integral for eta function for positive real numbers is

$$\eta(x) = \frac{1}{\Gamma(x)} \int_0^{\infty} \frac{t^{x-1}}{e^t + 1} dt, x > 0. \tag{6}$$

Eta function for complex variable S is defined as [22],

$$\eta(s) = \frac{1}{1^s} - \frac{1}{2^s} + \frac{1}{3^s} - ..., \operatorname{Re} s > 0,$$

3

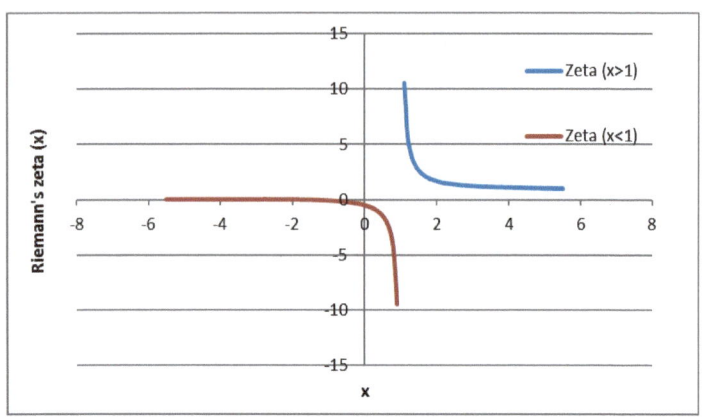

Figure 1. A plot of Riemann's zeta function.

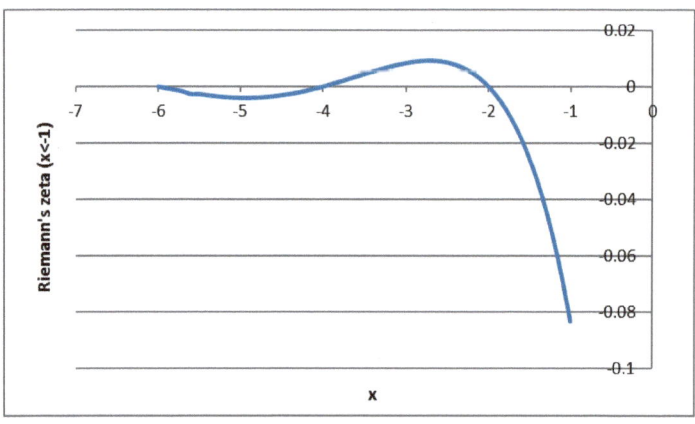

Figure 2. Riemann's zeta function for x<-1.

$$= \sum_{n=1}^{\infty} \frac{(-1)^{n-1}}{n^s}, \operatorname{Re} s > 0. \tag{7}$$

The integral for eta function is

$$\eta(s) = \frac{1}{\Gamma(s)} \int_0^\infty \frac{t^{s-1}}{e^t + 1} dt, \operatorname{Re} s > 0. \tag{8}$$

Eta of negative numbers is found using the functional equation

$$\eta(-s) = 2 \frac{1 - 2^{-s-1}}{1 - 2^{-s}} \pi^{-s-1} s \sin\left(\frac{\pi s}{2}\right) \Gamma(s) \eta(s+1).$$

Figure (3) and Table (2) present Dirichlet's eta function. Besides the authors mentioned, several others have contributed to the development of the zeta and eta functions. These include, P.L. Chebyshev, J. Hadamard, H.von Mangoldt, Ch. J. de la Vallee-Poussin etc. [24].

When we define zeta and eta functions to solve the harmonic series by replacing (x) with $(-x)$, i.e.,

$$\varsigma(-x) = \frac{1}{1^{-x}} + \frac{1}{2^{-x}} + \frac{1}{3^{-x}} + ..., \ x > 1$$

$$\eta(-x) = \frac{1}{1^{-x}} - \frac{1}{2^{-x}} + \frac{1}{3^{-x}} - ..., \ x > 0$$

it changes the nature of the series from harmonic to power series, i.e.,

$$\varsigma(-x) = 1^x + 2^x + 3^x + ..., \ x > 1$$

$$\eta(-x) = 1^x - 2^x + 3^x - ..., \ x > 0$$

respectively. Therefore the negative zeta and eta functions are defined as analytic continuation. There are several zeta and eta

5

functions, e.g., Huruwitz, Dadekind, arithmatic etc. [23]. This book extends the Euler-Riemann zeta and Dirichlet's eta functions to real negative and imaginary numbers without analytic continuation. In this book following nomenclature has been used:

Euler-Reimann's zeta function = $\varsigma(x)\,and\,\varsigma(s)$

Dirichlet's eta function = $\eta(x)\,and\,\eta(s)$

New zeta function = $\varsigma_a(c,x)$ = zeta_a

New eta function = $\eta_a(c,x)$ = eta _a

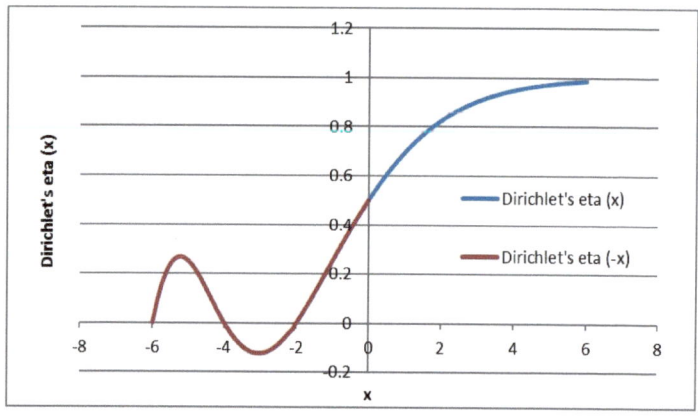

Figure 3. Plot of Dirichlet's eta function.

2

NEW ZETA FUNCTION WITHOUT ANALYTIC CONTINUATION

2.1 ς_a function for real negative numbers

A new zeta function, $\varsigma_a(-1,x)$ for real negative numbers which gives the sum of harmonic series may be defined as,

$$\varsigma_a(-1,x) = \sum_{n=1}^{\infty} \frac{1}{(-n)^x}, \quad x > 1$$

$$= \frac{1}{(-1)^x} + \frac{1}{(-2)^x} + \frac{1}{(-3)^x} + \dots, \quad x > 1$$

$$= \frac{1}{(-1)^x} \left[\frac{1}{1^x} + \frac{1}{2^x} + \frac{1}{3^x} + \dots \right], x > 1$$

$$\varsigma_a(-1,x) = \frac{1}{(-1)^x} \varsigma(x), x > 1. \tag{9}$$

Rectangular and polar graphs and values of $\varsigma_a(-1,x)$ are given in Figures (4-6) with certain values in Table (3).

It is seen from the graphs that the absolute value of $\varsigma_a(-1,x)$ is equal to that of $\varsigma(x)$. Since $\varsigma_a(c,x)$ for does not give the summation of the harmonic series, it has not been included in the present definition.

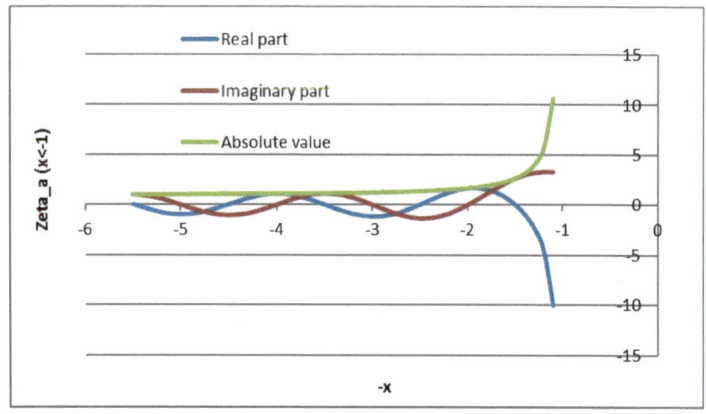

Figure 4. ς_a **function for** $x<-1$.

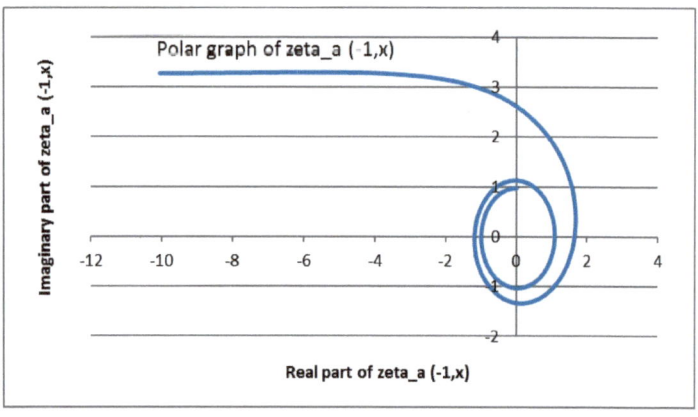

Figure 5. Polar graph of ς_a **function for** $x<-1$.

In general, for the harmonic series for zeta function, if n is multiplied by a constant c, we have,

$$\varsigma_a(c,x) = \frac{1}{(c)^x} + \frac{1}{(2c)^x} + \frac{1}{(3c)^x} + ..., c \neq 0, x > 1,$$

$$\varsigma_a(c,x) = \frac{1}{c^x}\varsigma(x), c \neq 0, x > 1. \tag{10}$$

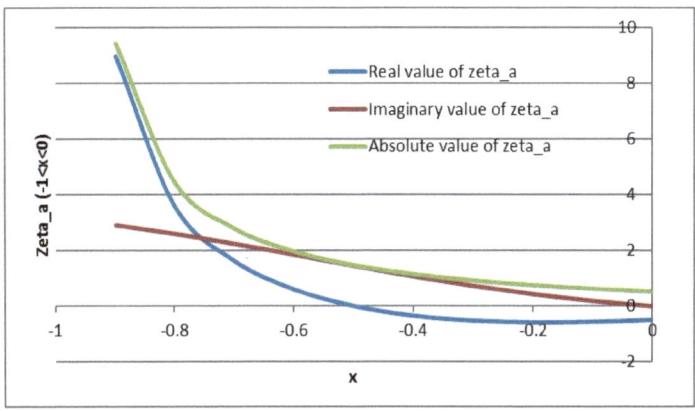

Figure 6. ς_a **function for -1<x<0.**

The integral of the function will be,

$$\varsigma_a(c,x) = \frac{1}{c^x \Gamma(x)}\int_0^\infty \frac{t^{x-1}}{e^t-1}dt,\ c \neq 0, x > 1. \tag{11}$$

The integral form of zeta function for the real negative axis Eqn. (9) will be,

$$\varsigma_a(-1,x) = \frac{1}{(-1)^x \Gamma(x)}\int_{-\infty}^0 \frac{(-t)^{x-1}}{e^{-t}-1}dt,\ x > 1. \tag{12}$$

Say for example, on the real negative axis zeta of (-1.5) will be given as:

$$\varsigma_a(-1,1.5) = \frac{1}{(-1)^{1.5}}\varsigma(1.5).$$

2.2 ς_a function for imaginary numbers

Substituting constant c in Eqn. (10) with imaginary number i for zeta of positive imaginary numbers,

$$\varsigma(i,x) = \frac{1}{i^x} + \frac{1}{(2i)^x} + \frac{1}{(3i)^x} + ..., x > 1$$

$$\varsigma_a(i,x) = \frac{1}{i^x}\varsigma(x), x > 1. \tag{13}$$

The integral form of the new zeta function for imaginary numbers will be,

$$\varsigma_a(i,x) = \frac{1}{i^x \Gamma(x)}\int_0^\infty \frac{t^{x-1}}{e^t - 1}dt, \ x > 1. \tag{14}$$

Similarly, for negative imaginary numbers substitute c with $-i$ in Eqn. (10),

$$\varsigma(-i,x) = \frac{1}{(-i)^x} + \frac{1}{(-2i)^x} + \frac{1}{(-3i)^x} + ..., x > 1,$$

$$\varsigma_a(-i,x) = \frac{1}{(-i)^x}\varsigma(x), x > 1. \tag{15}$$

The integral of ς_a for negative imaginary numbers will be,

$$\varsigma_a(-i,x) = \frac{1}{(-i)^x \Gamma(x)} \int_0^\infty \frac{t^{x-1}}{e^t-1}dt, \, x>1. \qquad (16)$$

Rectangular and polar charts of ς_a function of imaginary numbers are given in Figures (7-10).

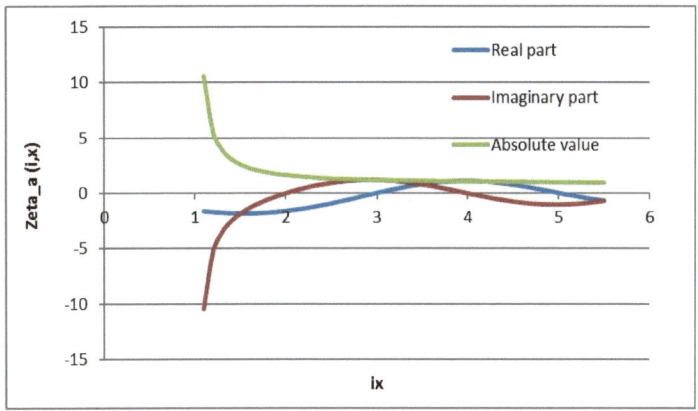

Figure 7. Graph of ς_a function for positive imaginary axis.

In general, for the harmonic series for zeta function, if n is multiplied by a constant c, we have,

$$\varsigma_a(c,x) = \frac{1}{(c)^x} + \frac{1}{(2c)^x} + \frac{1}{(3c)^x} + ..., c \neq 0, x>1,$$

$$\varsigma_a(c,x) = \frac{1}{c^x}\varsigma(x), c \neq 0, x>1.$$

11

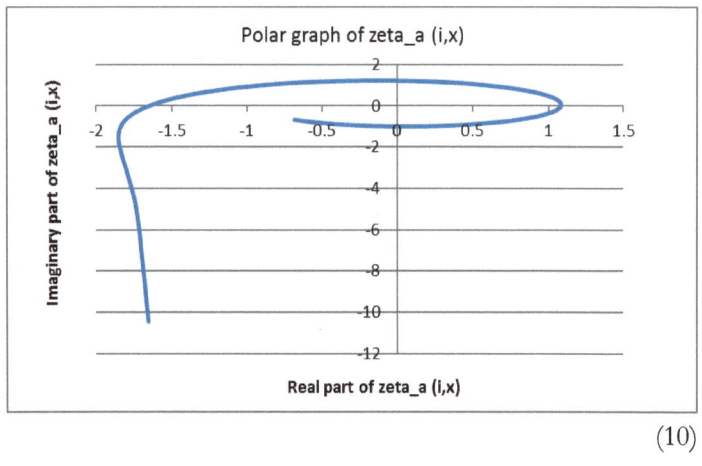

$$(10)$$

Figure 8. Polar graph of ς_a function for positive imaginary numbers.

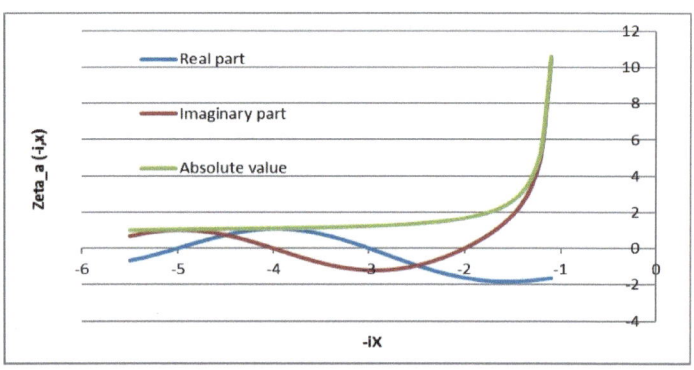

Figure 9. Graph of ς_a function for negative imaginary

2.3 Multi- and fractional ς_a functions

If in Eqn. (10), c is a mutiple or fraction of n (say $c = 2$ or 0.5), we can have multi- or fractional zeta functions. For double zeta function ($c = 2$), the series may be wrtten as,**axis.**

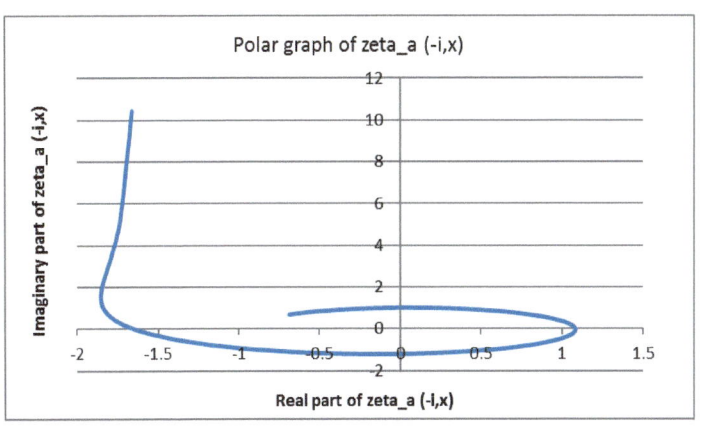

Figure 10. Polar graph of ς_a function for negative imaginary numbers.

$$\varsigma_a(2,x) = \frac{1}{2^x} + \frac{1}{4^x} + \frac{1}{6^x} + ...$$

$$\varsigma_a(2,x) = \frac{1}{2^x}\varsigma(x)$$

For $c = 0.5$, half zeta function, the series (Eqn. 10) will be,

$$\varsigma_a(0.5,x) = \frac{1}{0.5^x} + \frac{1}{1^x} + \frac{1}{1.5^x} + ...$$

$$\varsigma_a(0.5,x) = \frac{1}{0.5^x}\varsigma(x)$$

Fig. 11 and Table 4 give some multi- and fractional values of zeta_a function.

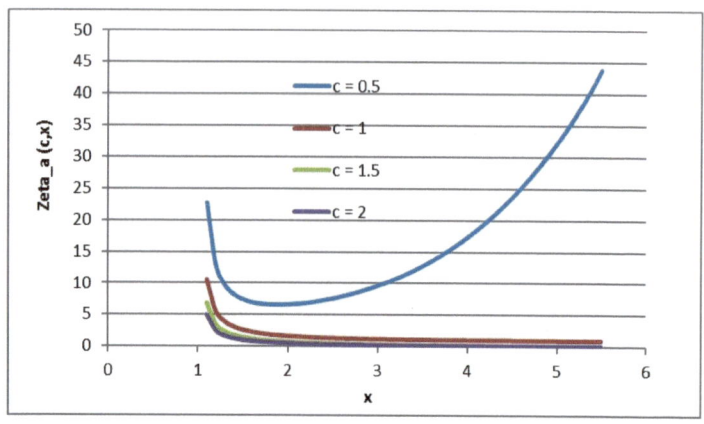

Figure 11. Plot for multi-ς_a and fractional ς_a functions.

3

NEW ETA FUNCTION WITHOUT ANALYTICAL CONTINUATION

3.1 η_a function for real negative numbers

A new eta function, η_a , for real negative numbers, which gives the sum of alternating harmonic series may be defined as,

$$\eta_a(-1,x) = \sum_{n=1}^{\infty} \frac{(-1)^{n-1}}{(-n)^x}, \quad x > 1$$

$$= \frac{1}{(-1)^x} - \frac{1}{(-2)^x} + \frac{1}{(-3)^x} - \dots, \quad x > 1$$

$$= \frac{1}{(-1)^x}\left[\frac{1}{1^x} - \frac{1}{2^x} + \frac{1}{3^x} - \dots\right], x > 1$$

$$\eta_a(-1,x) = \frac{1}{(-1)^x}\eta(x), x > 1. \tag{17}$$

In general, η_a , may be defined as,

$$\eta_a(c,x) = \frac{1}{(c)^x} - \frac{1}{(2c)^x} + \frac{1}{(3c)^x} - \dots, x > 0, c \neq 0$$

$$\eta_a(c,x) = \frac{1}{c^s}\eta(x), x > 0, c \neq 0, \tag{18}$$

where, $c = -1$ for eta function of real negative numbers, $c = i$ for positive imaginary numbers, and $-i$ for negative imaginary numbers.

Integral for η_a function will be,

$$\eta(c,x) = \frac{1}{c^x\Gamma(x)}\int_0^\infty \frac{t^{x-1}}{e^t+1}dt, x > 0, c \neq 0. \tag{19}$$

Integrals of the η_a function for negative real numbers is given below;

$$\eta(-1,x) = \frac{1}{(-1)^x\Gamma(x)}\int_0^\infty \frac{t^{x-1}}{e^t+1}dt, x > 0 \tag{20}$$

$$\eta(-1,x) = \frac{1}{(-1)^x\Gamma(x)}\int_{-\infty}^0 \frac{(-t)^{x-1}}{e^{-t}+1}dt, x > 0. \tag{21}$$

Rectangular and polar graphs for η_a function for real negative numbers are given in Figures (12,13) and Table 5.

3.2 η_a function for imaginary numbers

Substituting $c = i$ or $-i$, in Eqn. (17), η_a of imaginary numbers will be given as,

$$\eta_a(i,x) = \frac{1}{i^x} - \frac{1}{(2i)^x} + \frac{1}{(3i)^x} - \dots, x > 0$$

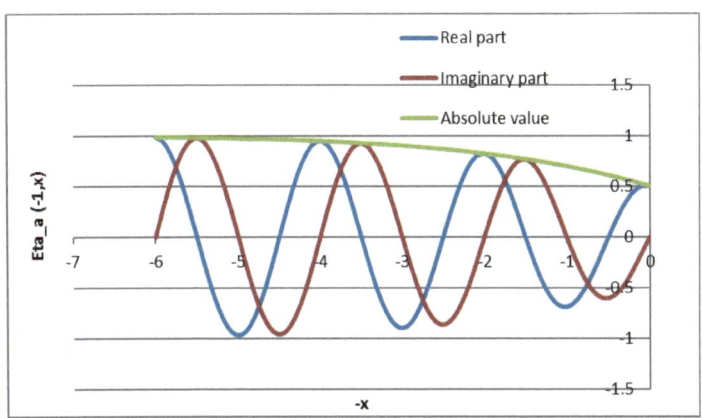

Figure 12. Plot of η_a function for real negative axis.

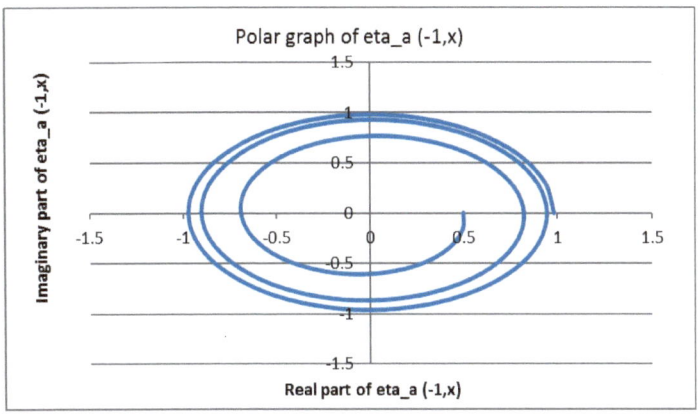

Figure 13. Polar graph for η_a function for real negative numbers.

$$\eta_a(i,x) = \frac{1}{i^x}\eta(x), x > 0. \tag{22}$$

$$\eta_a(-i,x) = \frac{1}{(-i)^x} - \frac{1}{(-2i)^x} + \frac{1}{(-3i)^x} - ..., x > 0$$

$$\eta_a(-i,x) = \frac{1}{(-i)^x}\eta(x), x > 0. \tag{23}$$

Recatangular and polar graphs for η_a function for imaginary numbers are given in Figures (14-17).

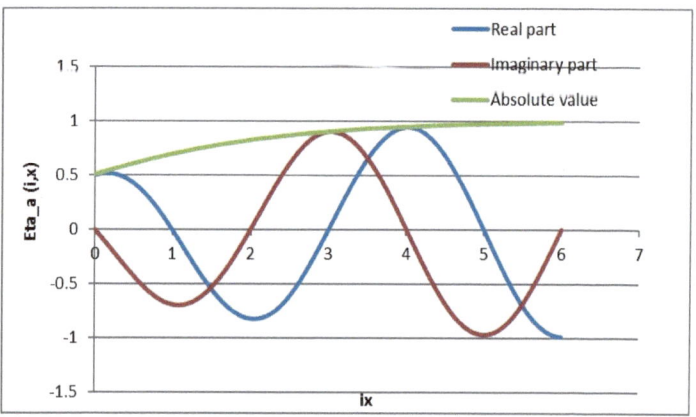

Figure 14. η_a **function for positive imaginary numbers.**

3.3 Multi-η_a and fractional η_a functions

If c is an integer or a fraction in Eqn. (18), multi-eta and fractional $\eta_a(c,x)$ functions may be defined.

Double eta function ($c = 2$),

$$\eta_a(2,x) = \frac{1}{2^x} - \frac{1}{4^x} + \frac{1}{6^x} - \ldots$$

$$\eta_a(2,x) = \frac{1}{2^x}\eta(x).$$

Fractional eta function (half eta function, $c = 0.5$)

$$\eta_a(0.5,x) = \frac{1}{0.5^x} - \frac{1}{1^x} + \frac{1}{1.5^x} - \ldots$$

$$\eta_a(0.5,x) = \frac{1}{0.5^x}\eta(x).$$

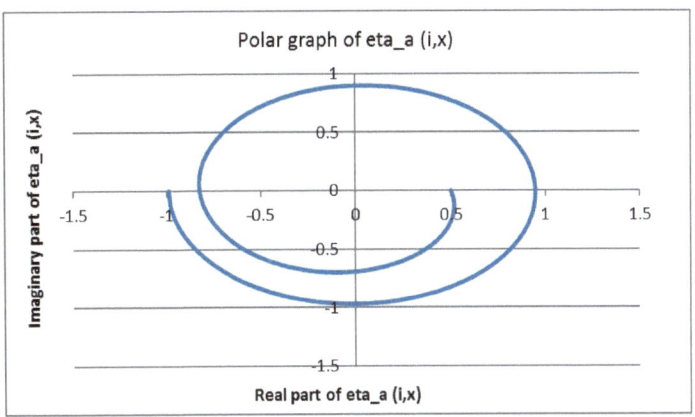

Figure 15. Polar graph of η_a function for positive imaginary numbers.

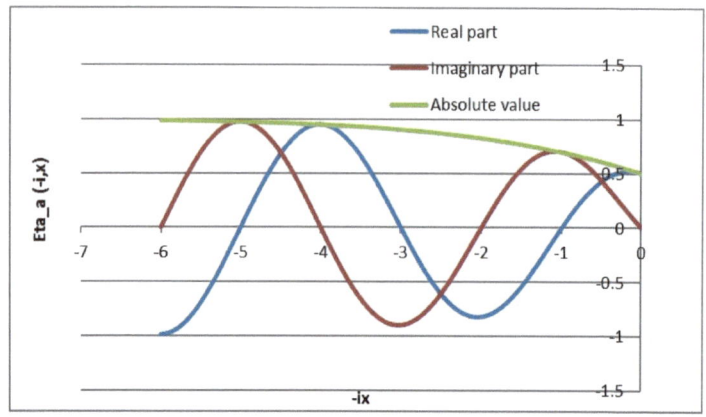

Figure 16. η_a **function for negative imaginary numbers.**

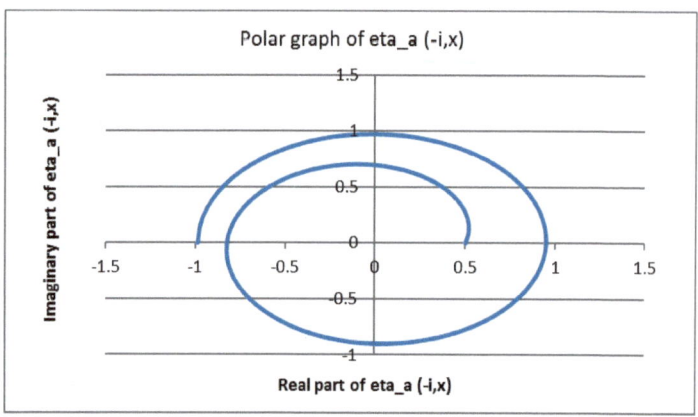

Figure 17. Polar graph of η_a **function for negative imaginary numbers.**

$$\eta_a(0.5, x) = \frac{1}{0.5^x}\eta(x).$$

If c is an integer or a fraction in Eqn. (18), multi-eta and fractional $\eta_a(c,x)$ functions may be defined as,

Double eta function ($c = 2$),

$$\eta_a(2,x) = \frac{1}{2^x} - \frac{1}{4^x} + \frac{1}{6^x} - ....$$

$$\eta_a(2,x) = \frac{1}{2^x}\eta(x).$$

Fractional eta function (half eta function, $c = 0.5$)

$$\eta_a(0.5,x) = \frac{1}{0.5^x} - \frac{1}{1^x} + \frac{1}{1.5^x} - ....$$

$$\eta_a(0.5,x) = \frac{1}{0.5^x}\eta(x).$$

Graph and some values of multi-eta and fractional η_a functions are given in Figure (18) and Table 6.

Table (7) presents the values of Riemann's zeta and Dirichlet's eta functions, and the proposed new functions of some integers for a comparison. Since the multiplication factor $\frac{1}{(-1)^x}$ in the proposed functions for the real negative numbers is a complex number, the values of these functions for real negative numbers will be complex numbers. However, at odd negative integers these will be real negative values and at even negative integers these will be real positive values.

Values of Riemann's zeta, Dirichlet's eta and proposed ς_a and η_a functions for some imaginary values of x are given in Tables (8,9).

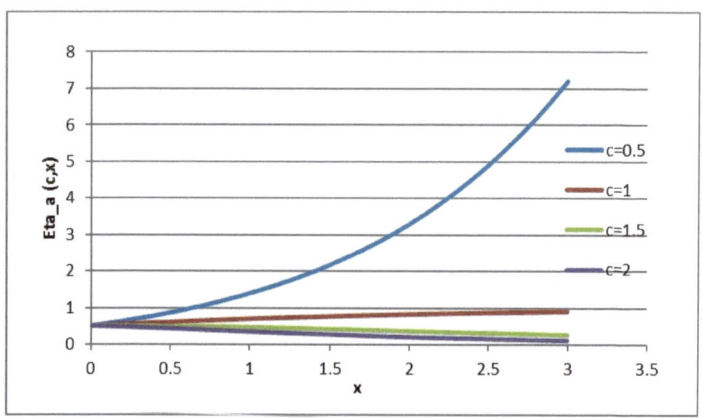

Figure 18. Plot of multi-η_a and fractional η_a functions.

3.4. Relation of ς_a and η_a functions with log function

The η_a function can be used to derive $\ln(1/2)$. Nicolas Mercator in the year 1668 gave a series [9],

$$\ln(1+x) = x - \frac{x^2}{2} + \frac{x^3}{3} - \dots, -1 < x \le 1.$$

For $x = 1$,

$$\ln(2) = x - \frac{1}{2} + \frac{1}{3} - \frac{1}{4} + \dots \quad = \eta(1).$$

Multiply both sides with -1,

$$-\ln(2) = -\eta(1) = \frac{1}{(-1)^1}\eta(1) = \eta_a(-1,1) = \ln(1/2). \quad (24)$$

Similarly,

$$\eta_a(i,1) = \frac{1}{i^1}\eta(1) = -i\ln(2),$$

and

$$\eta_a(-i,1) = \frac{1}{(-i)^1}\eta(1) = i\ln(2). \quad (25)$$

3.5 Relation between ς_a and η_a functions

Relation between new zeta and eta functions is also maintained for real negative numbers. Multiply both sides of Eqn. (2) with $\dfrac{1}{(-1)^x}$, we get,

$$\varsigma(-1,x) = \frac{\eta(-1,x)}{1-2^{1-x}}, x>0, x \neq 1. \quad (26)$$

The series of negative integers may also be summed (C_{neg}) using Ramanujan's summation. We know that the sum of positive integers (C_{pos}) by Ramanujan's summation is,

$$C_{pos} = -(1/12) = \varsigma(-1).$$

For negative real numbers,

$$
\begin{aligned}
C_{neg} &= & -1 & -2 & -3 & -4 & -5 & -6 & -... \\
4C_{neg} &= & & -4 & & -8 & & -12 & ... \\
-3C_{neg} &= & -1 & +2 & -3 & +4 & -5 & +6 & -... \\
&= & -(1 & -2 & +3 & -4 & +5 & -6 & ... \\
&= & -1/4 \\
C_{neg} &= & 1/12.
\end{aligned}
$$

In the present book, the original concepts of zeta and eta functions for real positive numbers on the basis of harmonic series and alternating harmonic series have been extended to real negative numbers in their natural form, i.e., without using analytic continuation. It is observed that earlier Euler-Reimann's zeta and Dirichlet's eta functions defined for real positive axis were extended to the real negative axis as analytic continuation. As a basic principle, zeta and eta functions for real negative axis should satisfy the sum of respective harmonic series on real negative axis. The earlier defined functions by analytic conuation do not have this property. The proposed ς_a and η_a functions satisfy this requirement. The negative and imaginary functions without analytical continuation may be defined as,

$$
\varsigma_a(c,x) = \frac{1}{c^x}\varsigma(x), x > 1, c \neq 0,
$$

and

$$
\eta_a(c,x) = \frac{1}{c^x}\eta(x), x > 0, c \neq 0,
$$

where $c = -1$ for negative, and i or $-i$ for imaginary numbers. The absolute values of the ς_a and η_a functions of negative and imaginary functions are equal to the values of Euler's zeta and Dirichlet's eta functions of real positive numbers respectively. The author has earlier defined the logarithms of negative and imaginary numbers, [15,16] and the factorials of negative and imaginary numbers [17]. The proposed ς_a and η_a functions give the values of the sum of zeta and eta series respectively on the real negative axis.

This is thus to conclude that the absolute values of the new proposed ς_a and η_a functions of negative and imaginary functions without analytic continuation are equal to the values of Euler's zeta and Dirichlet's eta functions of real positive numbers respectively

4

SOFTWARE USED FOR CALCULATIONS

Following softwares were used for computations and graph plotting:

1) Wolfram Alpha Examples: Zeta Functions (https://www.wolframalpha.com/examples/ZetaFunc tions.html)
2) Weisstein, Eric W. "Dirichlet Eta Function." From MathWorld--A Wolfram Web Resource. (http://mathworld.wolfram.com/DirichletEtaFunctio n.html)
3) Wolfram Alpha.com. Computation knowledge engine- Eta. http://www.wolframalpha.com/input/?i=eta%28-1%29
4) Gamma Function Evaluator - The Wolfram Functions Site (http://functions.wolfram.com/webMathematica/Fun ctionEvaluation.jsp?name=Gamma)
5) Wolfram Alpha.com. Infinite Series Calculator.http://www.wolframalpha.com
6) Wolfram Alpha Widgets: Series calculator- Free mathematics widget (2011). (www.wolfram.com/widgets/view.jsp?id=86ceba9f35 c96ebae137e44a36c7261a)
7) Wolfram Alpha Examples: Complex Numbers (https://www.wolframalpha.com/examples/Complex Numbers.html)
8) Wolfram Alpha, Math help Boards – Definite integral calculator.
9) MS-Excel

10) QaamGo Media GmbH, Hohenstaufenring 62, 50674 Köln, Germany Online convert.com

5

REFERENCES

[1] Ayoub, R. (1974) Euler and zeta function, American Mathematical Monthly, Mathematical Association of America. 81(10), 1067-1086. http://www.jstor.org/stable/2319041

[2] Borwein, P, Choi, S. Rooney, B. A. Weirathmueller, A. (Eds.) (2008) The Riemann Hypothesis: A Resource for the Afficionado and Virtuoso Alike, Springer, New York.

[3] Boyadzhev, K.N (2007) The Euler formula for $\varsigma(2n)$ - The Riemann zeta function and Bernoulli numbers, Mathematics Bonus Files for faculty and students, 1-7. http://www2.onu.edu/~m-caragiu.1/bonus_files/ZETA2N1.pdf

[4] Boyadzhev, K.N. Gadiyar H.G. and Padma, R. (2008) Alternating Euler sums at negative integers, arXiv: 0811.4437 [math.NT] http://arxiv.org/ftp/arxiv/papers/0811/0811.4437.pdf

[5] Davenport, H. (2000) Multiplicative Number Theory, Springer, New York,

[6] Dwilewicz R.J. and Minac, J. (2009) Values of Reimann's zeta function at integers, MAT². Materials Mathematics. Publicacio electronic de divulgacio del Department de Matematiques de la Universitat Autonoma de Barcelona. 6, 1-26. http://mat.uab.cat/matmat/pdf/PDFv2009/v2009n06.pdf.

[7]   Giuseppe, I. (2012) On some historical aspects of Riemann zeta function, 1. HAL archives-ouvertes, 1-7. https://hal.inria.fr/file/index/docid/828587/filename/RZF1.pdf

[8]   Gourdon X. and Sebah, P. (2004a) The Riemann zeta-function $\varsigma(s)$: generalities, Numbers, Constants and Computation.   1-8. http://numbers.computation.free/Constants/Miscellane ous/zetageneralities.pdf

[9]   Gourdon X. and Sebah, P. (2004b) The logarithmic constant: log   2 Numbers.computation.free.fr/Constants/constants.html . 1-24. Plouffe.fr/simon/articles/log2.pdf

[10] Milgram, M.S. (2013) Integral and series representations of Riemann's zeta function and Dirichlet's eta function and a medley of related results, Journal of Mathematics, article ID, 181724, 1-17.

[11] Odlyzko, (2015) Papers on zeros of the Riemann's zeta function   and   related   topics, www.dtc.umn.edu/~odlyzko/doc/zeta.html

[12] Olkonnen H. and Olkonnen, J.T. (2011) Log-time sampling of signals: zeta transform. Open Journal of Discrete   Mathematics,   1(2):   62-65. www.scirp.org/journal/PaperDownload.aspx?paperID= 5738

[13] Sierpinski, W. (1970) 250 Problems in Elementary Number Theory, Elsevier Publishing Co. Inc. New York.

[14] Srivastava, H.M and Junesang C. (2001) Series associated with the zeta and related functions. Springer Science & Business Media.

[15] Thukral A.K. and Parkash, O. (2014) A new approach for the logarithms of real negative numbers, Canadian Journal of Pure and Applied Sciences, 8(2): 2955–2961.

[16] Thukral, A.K. (2014) Logarithms of imaginary numbers in rectangular form: A new technique, Canadian Journal of Pure and Applied Sciences, 8(3): 3131–3137.

[17] Thukral, A.K. (2014) Factorials of real negative and imaginary numbers - A new perspective, SpringerPlus. 3: 658. 1-13. doi:10.1186/2193-1801-3-658 http://link.springer.com/article/10.1186%2F2193-1801-3-658#page-1

[18] Tyagi S. and Holm, C. (2007). A new integral representation for the Riemann zeta function. . arXiv:math-ph0703073v1. 7p.

[19] Weisstein, E.W. (2015a) Zeta function, From MathWorld- A Wolfram web resource. http://mathworld.wolfram.com/zeta function.html.

[20] Weisstein, E.W. (2015b) Dirichlet eta function, WolframMathworld- A Wolfram web resource. http://mathworld.wolfram.com/DirichletEtaFunction.html

[21] Wikipedia (2015a) Reimann zeta function,. https://en.wikipedia.org/wiki/Riemann_zeta_function

[22] Wikipedia (2015b) Dirichlet eta function, https://en.wikipedia.org/wiki/Dirichlet_eta_function

[23] Wikipedia (2015c) List of zeta functions. https://en.wikipedia.org/wiki/List_of_zeta_functions

[24] Wolfram Research (2015) Zeta Functions, Wolfram.com. http://functions.wolfram.com/ZetaFunctionsandPolylogarithms/Zeta/17/ShowAll.html

6

Tables

Table 1. Riemann's zeta values for real negative numbers.

x	Riemann's zeta	x	Riemann's zeta
0	-0.5		
-0.1	-0.41723	-3.1	0.00772
-0.2	-0.34967	-3.2	0.00701
-0.3	-0.29381	-3.3	0.00620
-0.4	-0.24717	-3.4	0.00534
-0.5	-0.20789	-3.5	0.00444
-0.6	-0.1746	-3.6	0.00352
-0.7	-0.14624	-3.7	0.00259
-0.8	-0.12199	-3.8	0.00169
-0.9	-0.10119	-3.9	0.00082
-1	-0.0833	-4	0
-1.1	-0.06799	-4.1	-0.00077
-1.2	-0.05479	-4.2	-0.00147
-1.3	-0.04346	-4.3	-0.00209
-1.4	-0.03377	-4.4	-0.00264
-1.5	-0.02549	-4.5	-0.00309
-1.6	-0.01845	-4.6	-0.00346
-1.7	-0.01251	-4.7	-0.00372
-1.8	-0.00752	-4.8	-0.00390
-1.9	-0.00339	-4.9	-0.00397
-2	0	-5	-0.00397
-2.1	0.00273	-5.1	-0.00387
-2.2	0.00487	-5.2	-0.00368

-2.3	0.00651	-5.3	-0.00341
-2.4	0.00771	-5.4	-0.00308
-2.5	0.00851	-5.5	-0.00267
-2.6	0.00898	-5.6	-0.00267
-2.7	0.00915	-5.7	-0.00170
-2.8	0.00908	-5.8	-0.00116
-2.9	0.00879	-5.9	-0.00059
-3	0.00833	-6	0

Table 2. Dirichlet's eta values for real negative numbers.

x	Dirichlet's eta	x	Dirichlet's eta
0	0.5		
-0.1	0.47712	-3.1	-0.12482
-0.2	0.453656	-3.2	-0.12186
-0.3	0.42964	-3.3	-0.11609
-0.4	0.405108	-3.4	-0.10748
-0.5	0.380105	-3.5	-0.09604
-0.6	0.35468	-3.6	-0.08184
-0.7	0.328889	-3.7	-0.06496
-0.8	0.302797	-3.8	-0.04555
-0.9	0.276474	-3.9	-0.02381
-1	0.25	-4	0
-1.1	0.223461	-4.1	0.02557
-1.2	0.196953	-4.2	0.05251
-1.3	0.170578	-4.3	0.08039
-1.4	0.144448	-4.4	0.10869
-1.5	0.118681	-4.5	0.13682
-1.6	0.093405	-4.6	0.16412
-1.7	0.068754	-4.7	0.18990
-1.8	0.04487	-4.8	0.21337
-1.9	0.021901	-4.9	0.23369
-2	0	-5	0.25
-2.1	-0.02674	-5.1	0.26137
-2.2	-0.03996	-5.2	0.26689
-2.3	-0.05769	-5.3	0.26562
-2.4	-0.07371	-5.4	0.25665
-2.5	-0.0878	-5.5	0.23912
-2.6	-0.09994	-5.6	0.21223
-2.7	-0.10984	-5.7	0.17532
-2.8	-0.1174	-5.8	0.12783

| -2.9 | -0.1225 | -5.9 | 0.06943 |
| -3 | -0.125 | -6 | 0 |

Table 3. New zeta_a function values for real negative numbers.

	Zeta function values		
-x	Real	Imaginary	Abs. value
-1.1	-10.06640	3.27075	10.58440
-1.2	-4.52368	3.28664	5.59158
-1.3	-2.31114	3.18101	3.93195
-1.4	-0.95967	2.95355	3.10555
-1.5	0	2.61238	2.61238
-1.6	0.70634	2.17389	2.28577
-1.7	1.20748	1.66195	2.05429
-1.8	1.52275	1.10634	1.88223
-1.9	1.66411	0.54070	1.74975
-2	1.64493	0	1.64493
-2.1	1.48385	-0.48213	1.56022
-2.2	1.20587	-0.87612	1.49054
-2.3	0.84195	-1.15885	1.43242
-2.4	0.42747	-1.31563	1.38334
-2.5	0	-1.34149	1.34149
-2.6	-0.40342	-1.24159	1.30548
-2.7	-0.74899	-1.0309	1.27426
-2.8	-1.00887	-0.73299	1.24703
-2.9	-1.16327	-0.37797	1.22313
-3	-1.20206	0	1.20206
-3.1	-1.12546	0.36568	1.18338
-3.2	-0.94394	0.68581	1.16677
-3.3	-0.67709	0.93193	1.15194
-3.4	-0.35187	1.08293	1.13866
-3.5	0	1.12673	1.12673
-3.6	0.34486	1.06137	1.11599
-3.7	0.65026	0.89500	1.10629
-3.8	0.88790	0.6451	1.09751

-3.9	1.03622	0.33668	1.08955
-4	1.08232	0	1.08232
-4.1	1.02200	-0.33207	1.07460
-4.2	0.86322	-0.62717	1.06700
-4.3	0.62270	-0.85708	1.05941
-4.4	0.32503	-1.00034	1.05181
-4.5	0	-1.04422	1.04422
-4.6	-0.32034	-0.98589	1.03662
-4.7	-0.60485	-0.83251	1.02903
-4.8	-0.82636	-0.60039	1.02143
-4.9	-0.96422	-0.31329	1.01384
-5	-1.00625	0	1.00624
-5.1	-0.94978	0.30860	0.99865
-5.2	-0.80178	0.58252	0.99105
-5.3	-0.57807	0.79563	0.98346
-5.4	-0.30156	0.92810	0.97586
-5.5	0	0.96827	0.96827

Table 4. Fractional and multi-zeta function.

x	$C = 0.5$	$C = 1$	$C = 1.5$	$C = 2$
		Zeta function values		
1.1	22.68816	10.58440	6.77588	4.93779
1.2	12.84608	5.59158	3.43736	2.43387
1.3	9.68159	3.93195	2.32107	1.59686
1.4	8.19559	3.10555	1.76039	1.17678
1.5	7.38892	2.61238	1.42200	0.92361
1.6	6.92915	2.28577	1.19477	0.75402
1.7	6.67440	2.05429	1.03111	0.63228
1.8	6.55430	1.88223	0.90721	0.54052
1.9	6.53029	1.74975	0.80984	0.46883
2	6.57972	1.64493	0.73108	0.41123
2.1	6.68881	1.56022	0.66587	0.36393
2.2	6.84872	1.49054	0.61086	0.32439
2.3	7.05406	1.43242	0.56371	0.29087
2.4	7.30131	1.38334	0.52276	0.26209
2.5	7.58861	1.34149	0.48681	0.23714
2.6	7.91495	1.30548	0.45491	0.21532
2.7	8.28016	1.27426	0.42639	0.19610
2.8	8.68482	1.24703	0.40070	0.17905
2.9	9.12976	1.22313	0.37740	0.16386
3	9.61648	1.20206	0.35616	0.15025
3.1	10.14652	1.18338	0.33669	0.13801
3.2	10.72213	1.16677	0.31878	0.12696
3.3	11.34564	1.15194	0.30222	0.11695
3.4	12.01977	1.13866	0.28686	0.10786
3.5	12.74749	1.12673	0.27258	0.09959
3.6	13.5322	1.11599	0.25925	0.09203
3.7	14.37739	1.10629	0.24679	0.08512
3.8	15.28701	1.09751	0.23510	0.07879

3.9	16.26538	1.08955	0.22412	0.07298
4	17.31712	1.08232	0.21379	0.06764
4.1	18.42770	1.07460	0.20383	0.06266
4.2	19.61073	1.06700	0.19435	0.05805
4.3	20.86865	1.05941	0.18529	0.05378
4.4	22.20612	1.05181	0.17666	0.04982
4.5	23.62808	1.04422	0.16841	0.04614
4.6	25.13976	1.03662	0.16054	0.04274
4.7	26.74671	1.02903	0.15303	0.03959
4.8	28.45484	1.02143	0.14587	0.03666
4.9	30.27038	1.01384	0.13903	0.03395
5	32.19995	1.00624	0.13251	0.03144
5.1	34.25056	0.99865	0.12628	0.02911
5.2	36.42967	0.99105	0.12034	0.02696
5.3	38.74513	0.98346	0.11467	0.02496
5.4	41.20531	0.97586	0.10926	0.02311
5.5	43.81905	0.96827	0.10411	0.02139

Table 5. New eta_a function values for some real negative numbers.

| -x | Eta function values | | |
	Real	Imaginary	Abs. value
0	0.5	0	0.5
-0.1	0.49670	-0.16139	0.52227
-0.2	0.44003	-0.31970	0.54391
-0.3	0.33204	-0.45702	0.56490
-0.4	0.18084	-0.55659	0.58523
-0.5	0	-0.60490	0.60489
-0.6	-0.19279	-0.59336	0.62389
-0.7	-0.37748	-0.51956	0.64220
-0.8	-0.53383	-0.38785	0.65985
-0.9	-0.64371	-0.20915	0.67683
-1	-0.69315	0	0.69314
-1.1	-0.67412	0.21903	0.70880
-1.2	-0.58559	0.42545	0.72382
-1.3	-0.43391	0.59722	0.73821
-1.4	-0.23238	0.71517	0.75198
-1.5	0	0.76514	0.76514
-1.6	0.24033	0.73965	0.77772
-1.7	0.46418	0.63890	0.78972
-1.8	0.64816	0.47091	0.80117
-1.9	0.77233	0.25094	0.81208
-2	0.82246	0	0.82246
-2.1	0.79161	-0.25721	0.83235
-2.2	0.68098	-0.49477	0.84174
-2.3	0.50001	-0.68821	0.85067
-2.4	0.26549	-0.81710	0.85915
-2.5	0	-0.86720	0.86720
-2.6	-0.27034	-0.83201	0.87483
-2.7	-0.51846	-0.71360	0.88206

-2.8	-0.71915	-0.52249	0.88891
-2.9	-0.85158	-0.27669	0.89540
-3	-0.90154	0	0.90154
-3.1	-0.86294	0.28038	0.90735
-3.2	-0.73850	0.53655	0.91284
-3.3	-0.53960	0.74269	0.91802
-3.4	-0.28520	0.87775	0.92292
-3.5	0	0.92755	0.92755
-3.6	0.28797	0.88630	0.93192
-3.7	0.55018	0.75727	0.93603
-3.8	0.76041	0.55247	0.93992
-3.9	0.89740	0.29158	0.94358
-4	0.94703	0	0.94703
-4.1	0.90377	-0.29365	0.95028
-4.2	0.77127	-0.56036	0.95334
-4.3	0.56205	-0.77360	0.95622
-4.4	0.29632	-0.91200	0.95893
-4.5	0	-0.96148	0.96148
-4.6	-0.29786	-0.91671	0.96388
-4.7	-0.56788	-0.78162	0.96613
-4.8	-0.78334	-0.56913	0.96825
-4.9	-0.92276	-0.29982	0.97024
-5	-0.97212	0	0.97211
-5.1	-0.92621	0.30094	0.97387
-5.2	-0.78946	0.57357	0.97582
-5.3	-0.57431	0.79047	0.97707
-5.4	-0.30238	0.93063	0.97853
-5.5	0	0.97989	0.97989
-5.6	0.30320	0.93315	0.98117
-5.7	0.57742	0.79476	0.98237
-5.8	0.79567	0.57809	0.98350
-5.9	0.93637	0.30424	0.98456
-6	0.98555	0	0.98555

Table 6. Fractional and multi-eta functions.

x	$C = 0.5$	$C = 1$	$C = 1.5$	$C = 2$
	Eta function values			
0	0.5	0.5	0.5	0.5
0.1	0.55975	0.52227	0.50151	0.48729
0.2	0.62478	0.54391	0.50154	0.47350
0.3	0.69547	0.56490	0.50020	0.45884
0.4	0.77222	0.58523	0.49761	0.44352
0.5	0.85545	0.60489	0.49389	0.42772
0.6	0.94564	0.62389	0.48916	0.41161
0.7	1.04327	0.64220	0.48351	0.39532
0.8	1.14887	0.65985	0.47706	0.37898
0.9	1.26301	0.67683	0.46989	0.36270
1	1.38629	0.69314	0.46209	0.34657
1.1	1.51936	0.70880	0.45376	0.33067
1.2	1.66291	0.72382	0.44496	0.31506
1.3	1.81769	0.73821	0.43577	0.29980
1.4	1.98449	0.75198	0.42626	0.28494
1.5	2.16416	0.76514	0.41649	0.27052
1.6	2.35761	0.77772	0.40651	0.25655
1.7	2.56582	0.78972	0.39638	0.24306
1.8	2.78984	0.80117	0.38615	0.23007
1.9	3.03079	0.81208	0.37586	0.21759
2	3.28986	0.82246	0.36554	0.20561
2.1	3.56836	0.83235	0.35523	0.19415
2.2	3.86765	0.84174	0.34497	0.18319
2.3	4.18921	0.85067	0.33477	0.17274
2.4	4.53464	0.85915	0.32467	0.16277
2.5	4.90562	0.86720	0.31469	0.15330
2.6	5.30398	0.87483	0.30485	0.14429
2.7	5.73166	0.88206	0.29515	0.13574
2.8	6.19077	0.88891	0.28563	0.12763

2.9	6.68352	0.89540	0.27628	0.11995
3	7.21234	0.90154	0.26712	0.11269
3.1	7.77978	0.90735	0.25816	0.10582
3.2	8.38862	0.91284	0.24940	0.09933
3.3	9.04179	0.91802	0.24085	0.09320
3.4	9.742470	0.92292	0.23251	0.08743
3.5	10.49408	0.92755	0.22439	0.08198
3.6	11.30021	0.93192	0.21649	0.07685
3.7	12.16479	0.93603	0.20881	0.07202
3.8	13.09199	0.93992	0.20134	0.06748
3.9	14.08631	0.94358	0.19409	0.06320
4	15.15248	0.94703	0.18706	0.05918
4.1	16.29579	0.95028	0.18025	0.05541
4.2	17.52164	0.95334	0.17364	0.05187
4.3	18.83598	0.95622	0.16725	0.04854
4.4	20.24514	0.95893	0.16106	0.04542
4.5	21.7559	0.96148	0.15507	0.04249
4.6	23.37555	0.96388	0.14928	0.03974
4.7	25.11191	0.96613	0.14368	0.03717
4.8	26.97333	0.96825	0.13827	0.03475
4.9	28.96875	0.97024	0.13305	0.03249
5	31.10781	0.97211	0.12801	0.03037
5.1	33.40082	0.97387	0.12315	0.02839
5.2	35.86979	0.97582	0.11849	0.02654
5.3	38.49353	0.97707	0.11393	0.02480
5.4	41.31778	0.97853	0.10956	0.02317
5.5	44.34508	0.97989	0.10536	0.02165
5.6	47.58996	0.98117	0.10130	0.02022
5.7	51.06809	0.98237	0.09740	0.01889
5.8	54.79621	0.98350	0.09363	0.01765
5.9	58.79219	0.98456	0.09001	0.01648
6	63.07520	0.98555	0.08652	0.01539

Table 7. Values of Riemann's zeta, Dirichlet's eta and proposed ς_a and η_a functions for some integer values of x.

x	Riemann's zeta (x)	Present zeta (c,x)	Dirichlet's eta (x)	Present eta (c,x)
	x (Positive real numbers)			
0	-0.5	--	0.5	0.5
1	Inf	Inf	0.6931	0.6931
2	1.6449	1.6449	0.8224	0.8224
3	1.2020	1.2020	0.9015	0.9015
4	1.0823	1.0823	0.9470	0.9470
5	1.0369	1.0369	0.9721	0.9721
6	1.0173	1.0173	0.9855	0.9855
7	1.0083	1.0083	0.9955	0.9955
8	1.0040	1.0040	0.9620	0.9620
9	1.0020	1.0020	0.9980	0.9980
10	1.0009	1.0009	0.9990	0.9990
	$-x$ (Negative real numbers)			
-1	-0.0833	--	0.25	-0.6931
-2	0	1.6449	0	0.8224
-3	0.0083	-1.2020	-0.125	-0.9015
-4	0	1.0823	0	0.9470
-5	-0.0039	-1.0369	0.25	-0.9721
-6	0	1.0173	0	0.9855
-7	0.0041	-1.0083	-1.0625	-0.9955
-8	0	1.0040	0	0.962
-9	-0.0075	-1.0020	7.75	-0.9980
-10	0	1.0009	0	0.9990

"—" not defined

Table 8. Values of Riemann's zeta and proposed ς_a function for some imaginary values of x.

x	Riemann's zeta (ix)	Present zeta (ς,x)
	ix (Positive imaginary numbers)	
i	0.0033-0.4181 i	--
2 i	0.3147-0.2316 i	-1.6449
3 i	0.4392-0.0364 i	1.2020 i
4 i	0.5277+0.1342 i	1.0823
5 i	0.6330+0.2906 i	-1.0369 i
6 i	0.7862+0.4264 i	-1.0173
7 i	1.0068+0.5116 i	1.0083 i
8 i	1.2904+0.4911 i	1.0040
9 i	1.5821+0.2980 i	-1.0020 i
10 i	1.7564-0.1015 i	-1.0009
	$-ix$ (Negative imaginary numbers)	
$-i$	0.0033+0.4181 i	--
-2 i	0.3147+0.2316 i	-1.6449
-3 i	0.4392+0.0364 i	-1.2020 i
-4 i	0.5277-0.1342 i	1.0823
-5 i	0.6330-0.2906 i	1.0369 i
-6 i	0.7862-0.4264 i	-1.0173
-7 i	1.0068-0.5116 i	-1.0083 i
-8 i	1.2904-0.4911 i	1.0040
-9 i	1.5821-0.2980 i	1.0020 i
-10 i	1.7564+0.1015 i	-1.0009

"—" not defined

Table 9. Values of Dirichlet's eta and proposed η_a function for some imaginary values of x.

x	Dirichlet's eta (ix)	Present eta (c,x)
	ix (Positive imaginary numbers)	
i	0.5325+0.2293 i	-0.6931 i
2 i	0.6547+0.4720 i	-0.8224
3 i	0.9308+0.6953 i	0.9015 i
4 i	1.4152+0.7653 i	0.9470
5 i	2.0184+0.4384 i	-0.9721 i
6 i	2.3383-0.4628 i	-0.9855
7 i	1.7399-1.6248 i	0.9955 i
8 i	0.0419-1.9719 i	0.962
9 i	-1.5522-0.4393 i	-0.9980 i
10 i	-0.9211+2.1815 i	-0.9990
	$-ix$ (Negative imaginary numbers)	
$-i$	0.5325-0.2293 i	0.6931 i
-2 i	0.6547-0.4720 i	-0.8224
-3 i	0.9308-0.6953 i	-0.9015 i
-4 i	1.4152-0.7653 i	0.9470
-5 i	2.0184-0.4384 i	0.9721 i
-6 i	2.3383+0.4628 i	-0.9855
-7 i	1.7399+1.6248 i	-0.9955 i
-8 i	0.0419+1.9719 i	0.9620
-9 i	-1.5522+0.4393 i	0.9980 i
-10 i	-0.9211-2.1815 i	-0.9990

Ashwani K Thukral

Zeta and Eta Functions
A new hypothesis

Euler-Riemann's zeta and Dirichlet's eta functions are defined for real negative numbers as analytic continuation. In the present book the author defines new series for zeta(ς_a) and eta (η_a) functions, for real negative and imaginary numbers without analytic continuation. The new zeta and eta functions maintain the character of harmonic series and alternating harmonic series respectively for real negative numbers, as for real positive numbers. Zeta and eta functions have also been defined for multiple- and fractional harmonic series.

Dr. A.K.Thukral was born in 1953 at Amritsar (Punjab), India, and got his Ph.D. (1980) from BITS, Pilani, India. He was a Postdoctoral fellow (1984-86) at the Lomonosov Moscow State University, Moscow. He has been on the faculty of Guru Nanak Dev University since 1980. Dr. Thukral specialises in Environmental Sciences. He has supervised 15 Ph.D. students, published more than 125 research papers, edited 2 books and attended several conferences. He has held various academic positions, such as Head of the Department, Dean of the Faculty, Dean of Student's Welfare and Director of Research. He is a crusader of integrated and interdisciplinary education and research.

Ashwani K Thukral